Science Around Us

Dark and Light

Sally Hewitt

Chrysalis Children's Books

First published in the UK in 2003 by

(✿) Chrysalis Children's Books

64 Brewery Road, London N7 9NT

ISBN 1 84138 721 5

British Library Cataloguing in Publication Data
for this book is available from the British Library.

A BELITHA BOOK

Editorial manager: Joyce Bentley
Project editor: Clare Weaver
Designer: Wladek Szechter
Picture researcher: Aline Morley
Consultant: Helen Walters

Printed in Hong Kong

10 9 8 7 6 5 4 3 2 1

Words in **bold** can be found in Words to remember on page 30.

Picture credits
Cover main ; Guy Motil/Corbis
Inset (L-R) ; Loisjoy Thurston/Bubbles, Jennie Woodcock/Bubbles, R.W. Jones/Corbis, Loisjoy Thurston/Bubbles, Back cover ; Richard Ransier/Corbis © Bubbles P4 Jennie Woodcock, P5 Loisjoy Thurston, P12 Chris Rout, P13 Loisjoy Thurston, P20 Loisjoy Thurston, P22 Ian West, P25 Loisjoy Thurston, P26 Angela Hampton, P27 Ian West © Corbis P1 Richard Ransier, P5 Paul A. Souders, P6 R.W Jones, P7 Jennie Woodcock/Reflections Photolibrary, P8 George Shelley, P9 Guy Motil, P11 Richard Ransier, P14 PBNJ Productions, P15 Wayne Bennett, P16 Bohemian Nomad Picturemakers, P17 Jay Dickman, P18 Royalty-Free, P19 Pat Doyle, P21 (L) Craig Aurness, P21 (R) Bohemian Picturemakers, P23 Picture Press, P24 Gina Minielli © Getty P10 Darrell Gulin

Contents

What is light?

The **Sun** gives us **light**. It is a giant ball of burning gas in space that shines so brightly it lights up our planet Earth.

Light lets us see **colours** and shapes and things moving all around us.

Daytime begins when the Sun rises in the morning.

Daytime ends when the Sun disappears at sunset.

The Sun shines during the day. Most people wake up and work or play during the day when it is light and they can see.

What is dark?

It is dark when there is no light shining. Night-time is dark because we cannot see the Sun.

At night we use electric lights to light up the dark.

Night-time begins when the Sun sets in the evening. It ends when the Sun rises in the morning.

We cannot see very well in the dark, so night is a good time to go to sleep.

Earth spins round in space. It is night when your part of the Earth faces away from the Sun.

Glowing light

When something gets very hot it **glows** and gives out light. The hot sun glows so brightly it lights up nine planets in space. A glowing fire only gives out a little light.

Hot flames send out flickering firelight.

The Moon does not glow because it is a ball of rock. The moonlight we see is really sunlight shining onto the Moon.

Moonlight is pale and does not give out any heat.

The Moon acts like a mirror. It **reflects** light from the sun back to Earth.

Rays and shadows

All kinds of light – sunlight, lamplight and candlelight – travel in straight lines called **rays**. Light cannot shine through solid things such as a wall or a tree, and it cannot go around corners.

On a sunny day you can see the rays of sunlight shining between the leaves and branches of trees.

Your shadow is the same shape as you, and it does all the same things that you do.

You are solid so light cannot shine through you. On a sunny day you make a dark shape where the sun cannot reach called a **shadow.**

We see with our eyes

We see the world around us when light shines into our eyes. We can only see with our eyes open. When we shut our eyes, we shut out the light and we cannot see.

Light goes into our eyes through the black holes called the **pupils**.

A camera **lens** is like an eye. The shutter is like an eyelid that blinks open and shut to let light in and out.

Light shines through the lens onto the film inside the camera. This takes a photograph of what you see through the view finder.

Your pupils open wide in the dark to let in as much light as possible. They get smaller in bright light.

Animal sight

Many animals see things in a different way to us. Nocturnal animals hunt for their food at night. Their eyes let them see much better in the dark than we can.

A cat hunts at night. Its eyes are so good at letting in light that they glow in the dark.

A hawk's eyes both face forward. This helps it to work out the exact position of a small animal on the ground.

A hawk flies high in the air and hunts for food on the ground. It has very sharp eyesight.
It spots the smallest movement far below and swoops down on its prey.

Rainbows

When it rains and the sun is shining at
the same time, turn your back to the Sun and
look up at the sky. You might see a **rainbow**.

We call sunlight 'white light' because it
doesn't seem to have any one colour.

White light is really made up of colours we cannot usually see. When sunlight shines through raindrops it bends and splits into seven colours and you see a rainbow.

You can make a rainbow by shining light through a piece of shaped glass called a **prism**.

The colours of the rainbow are called the **spectrum**. They are red, orange, yellow, green, blue, indigo and violet.

Seeing colour

Some animals can only see the world in black and white. Our eyes let us see colour. Colours give us information about our surroungings and help to make the world look beautiful and interesting.

We see different colours because of the way light bounces into our eyes. Apples look red because red light **reflects** (or bounces) off them into our eyes.

Cells in your eyes called cones send messages to your brain about the colours you see.

The coat of the black kitten absorbs all the colours of light. The coat of the white kitten reflects all the colours of light.

Things look black because they **absorb** all the colours of light. No colours reflect off black objects into our eyes. Things look white when all the colours reflect into our eyes.

See through and non-see through

Light shines through **transparent** materials like glass. We can see through any materials that let all the light shine through.

Glass is a good material for a toyshop because it lets you see the toys for sale.

Materials such as tissue paper or the dark glass in sunglasses are **translucent**. Translucent materials only let some light through.

Clouds are translucent. It is dull on a cloudy day because clouds only let some sunlight shine through.

Solid things like a wooden door are **opaque**. Light cannot shine through them and so you cannot see through them either.

Reflection

When you look at a mirror, you see a picture of yourself looking back at you. This picture is called a **reflection.**

Light bounces off the shiny, flat surface of the mirror straight back into your eyes and you see a clear reflection.

Your reflection in a mirror looks the wrong way round. Your right hand looks as though it is your left hand.

Light bounces off rippling water in all directions and makes the reflection look wobbly.

Still water in a lake or puddle is flat and shiny like a mirror. You can see a clear reflection of yourself, the sky and the trees and buildings around it.

Speeding *light*

Light travels faster than anything else in the universe. It speeds through air but it slows down when it shines through water and glass.

Light bends or **refracts** a little bit as it slows down. This makes you look bent when you are half in and half out of water.

24

Light travels very fast all the time. It never stops moving.

When light bends, it can play tricks with the way you see things.

A drinking straw seems to bend at the place where it enters the water, but it is not really bent. If you take the straw out of the glass, it looks straight again.

Bigger and smaller

Lenses are curved pieces of glass. They change the way you see things when you look through them. They make things look bigger or smaller and nearer or farther away.

Different shaped lenses in glasses help short-sighted and long-sighted people see more clearly.

Convex lenses are thicker in the middle. They make things look bigger. Concave lenses are thinner in the middle. They make things look smaller.

Powerful binoculars are good for bird-spotting and star-gazing. They make the birds and stars look bigger and nearer.

If you look through the narrow end of binoculars, the lenses inside them make things in the distance look nearer.

MAKE SHADOW PUPPETS

Shine light onto shapes and put on a shadow puppet show.

YOU WILL NEED:

- a sheet of dark card
- pencil
- scissors
- sticky tape
- 2 drinking straws
- a large sheet of thick card
- a sheet of tracing paper
- a torch

1 Draw a teddy and a dog onto the dark card. Cut out the shapes.

2 Tape a drinking straw onto the back of each character for handles.

3 Bend back a 10cm strip along one edge of the piece of thick card.

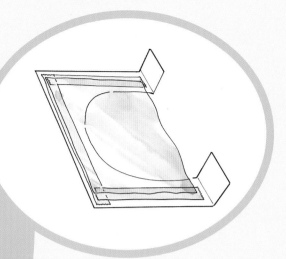

4 Cut a large hole in the card for the screen.

5 Tape the tracing paper across the back of the screen.

PUT ON A SHOW

1 Shine a bright torch at the screen from behind.

2 Move your puppets between the screen and the torch. Your friends will see the shadow puppets moving on the screen.

Words to remember

absorb To take in. Things look black because they absorb all the colours of light.

cells Tiny parts of living things. Each cell in your body has a job to do. Cells in your eyes let you see light and colour.

cloud A mist hanging in the sky, made up of millions of tiny drops of water hanging in groups in the sky. It is dull on a cloudy day because clouds only let some sunlight shine through,

colour Light is made up of seven colours: red, orange, yellow, green, blue, indigo and violet. We see colours when one or more of the colours of light bounces off an object and into our eyes.

glow When something gets so hot it gives out heat. The Sun, fire and a light bulb all glow.

lens A curved piece of glass or plastic that makes things look bigger or smaller and nearer or farther away. We see things when light shines through lenses inside our eyes.

light Lets us see things. The Sun shines so brightly that it lights up our planet Earth and lets us see by day. At night, we see by lamplight or candlelight.

opaque Solid things like a wall, a door or you. Light cannot shine through things that are opaque, so you cannot see through them.

prism A piece of glass shaped like a triangle. When white light shines through a prism it bends and splits into all the colours of the rainbow.

pupils The black circles in the middle of your eyes, which are really holes. Your pupils get bigger and smaller to let in and keep out light.

rainbow An arc of colour in the sky. We see a rainbow when sunlight bends as it shines through raindrops and splits into seven colours.

ray A straight line of light. Light travels very fast in rays. We can sometimes see rays of sunlight shining through clouds or through trees.

reflect When light bounces off a very shiny surface like a mirror. When you look in a mirror, you see a picture of yourself called a reflection.

refract When light bends a little as it shines through glass or water.

shadow A dark patch made when light cannot shine through something solid. Light cannot shine through you, so you make a shadow.

spectrum The seven colours (red, orange, yellow, green, blue, indigo, violet) seen when light shines through raindrops or a prism.

Sun A ball of burning gas. The Sun lights up our days and heats up our planet Earth.

translucent Things like a cloud or tissue paper that only let some light shine through them.

transparent Things like glass or water that let light shine through them. We can see through things that are transparent.

Index